STELLAR ATMOSPHERES
ALICIA SOMETIMES

ALICIA
SOMETIMES

STELLAR

ATMOSPHERES

BOOK 09
SERIES 5

CORDITE
BOOKS

First printed in 2024
by Cordite Publishing Inc.

We acknowledge the Boonwurrung and Wurrundjeri Woi Wurrung
peoples of the Eastern Kulin Nation, on whose land most of this
collection was written on.

PO Box 58
Castlemaine 3450
Victoria, Australia
cordite.org.au | corditebooks.org.au

National Library of Australia
Cataloguing-in-Publication:

   Sometimes, Alicia
   Stellar Atmospheres
   978-0-6457616-2-7 paperback
   I. Title.
   A821.3

Poetry set in Rabenau 10 / 15
Cover design by Zoë Sadokierski
Text design by Kent MacCarter and Zoë Sadokierski
Printed and bound by McPhersons in Maryborough, Victoria

10 9 8 7 6 5 4 3 2 1

Universal love to my wonderful family: Steve, Arlo and Jasper.

CONTENTS

RED GIANTS

SUPERNOVA

ACKNOWLEDGEMENTS

PREFACE

From the first moment I listened to stories about the night sky, I fell in love with astronomy and physics. Not just the stars or galaxies but also the why and how of it all.

What happened before the Big Bang? What are we made of? How will the universe end? I am not a scientist, just eternally curious and have made scientific research the basis for a great deal of my poetry. The cosmos is endlessly fascinating. And writeable.

I have been fortunate enough to work alongside inspiring scientists and have read and listened to many more. Scientists regularly and successfully use vivid storytelling and poetics, using metaphor to weave into their factual narrative. It can be an indispensable tool, helping us understand something like Newton's second law of thermodynamics or dark matter.

Sometimes, adding abstraction (poetry) on top of abstraction (difficult to understand scientific concepts) can feel like you're going on a side quest. In those splinters of time, I hope you stay with me.

Physicist Niels Bohr said, 'What is it that we humans depend on? We depend on our words. We are suspended in language. Our task is to communicate experience and ideas to others.'

With this, I am humbly grateful and thankful to be suspended in language with you.

INTRODUCTION

I feel a sense of delight at the idea of an artist surreptitiously working in a science lab. There is something mischievous, rambunctious, even anarchistic about it. The idea of intervention. I have always thought that the disciplines that exist under the broad umbrellas of science and art are in some ways artificial necessities for the organisation of various institutions. Of course, science and art embody different *ways* of knowing, of epistemological knowledge-making, but there are forms of art that bleed together with scientific practice more so than two disciplines thought of as sciences – consider the techniques used in optical microscopy and cinematography (both lens based practices), versus geology and biomedical science (rocks versus the messy stuff of humans and disease).

When poetry turns its reflective gaze onto astronomical phenomena, concepts and language, what emerges is a profound connection of science to the human condition, a way of experiencing scientific phenomena in ways that cannot be experienced through scientific perspectives alone. The subjective is acknowledged, unveiled, celebrated.

The poems in Sometimes's collection deftly transcend both spatial scales and time scales. From one line to another we careen across the universe. We fast-forward from the first picosecond of stuff forming in the universe to a Christmas card, millennia later. Her depiction of time and dynamism is visceral – things froth and whizz and quiver in a temporospatial-grammatical practice. We become aware of the minutiae of a life, and of a language, against both vast and infinitesimal phenomena in the universe.

Lives are writ large in *Stellar Atmospheres*. Biographical poems about female scientists scaffold the collection. These women's extensive contribution to astronomy, and the sacrifices they made for the sake of knowledge about the universe, leave you with a sense of the institutional and ethical impacts of overshadowing these scientists. These poems unfold with curiosity and respect rather than malice.

Nothing about this writing shies away from the science – Sometimes's deep knowledge of astrophysics and cosmology reflects her longstanding collaborations with some of Australia's leading scientists. The book is a laboratory of its own, at times taming, liberating or penetrating the science. It probes the linguistic choices that astronomy has made.

Instruments, biographies, bodies: celestial and human. Poems gather forms, shapes and grammar for phenomena that cannot be directly observed even when our senses are augmented by complex technoscientific systems.

'Astronomers rarely look up at the sky
Instruments detect invisible signals'

Scientists and engineers design the human perception of these signals using graphs, visualisations and sound, but Sometimes contributes a unique poetic-linguistic translation of astronomical phenomena. Nebulae and supernovae are transmitted to the reader as 'blue stencils in space' and 'brilliant dancers'. A red giant is evoked via Rothko:

'Now he stands before this ache of colour
in fluster, blushing deferential-cranberry'

Sometimes has written the celestial subjectively – or is it the subjective celestially? – and smears the lines between the practices of cosmology, poetry, and astrophysics. After reading this collection, it seems to me like poetry is as valid a method of interrogating the universe as any other.

—Andrea Rassell

NEBULA

First Three Minutes   for Steven Weinberg

there was an expansion. simultaneously. everywhere.
not at zero second. but close. after the first picoseconds.
hotter than the hottest star. blue hot. a nursery of brooding
hot. tempestuous bright. a Norse myth says, 'Earth had not
been, nor heaven above – but a yawning gap and grass
nowhere.' There were no Vikings, no lamp shades, no vanilla
ice cream or deep pools of lava
         but there was Lego. for life

in the first three minutes of the universe everything that would
be anything started to ferment. clump together. ready to sing
at a recital. began to develop lips to form the word *poem*. it all
moved towards a Christmas card that got people we don't know
back together. after this – one star dreamed of turning away
another because it needed time to shape clay. the universe
became a rogue gallery of jigsaws fighting for space and in softer
moments, mango juice squeezed from the heavens
         sparkled like iridescent jellyfish suits

next there was the first spoonful of the Cartwheel Galaxy, NGC
1365 with its gem-like barred spiral worlds, Jupiter's arresting
moons and Saturn's non-spherical Hyperion. pulsars were born
cramping the style of black holes and stealing limelight off
everything unseen. chambers of birth. in the first three minutes
of the universe hints of molecules began collecting so the word
         pokeweed could be part of this story

in the first three minutes
matter won over anti-matter
subatomic particles diffused
their presence amongst all spectra

as Rumi said

> *the atoms are dancing*
> *happy or miserable*

and they just kept on dancing

Caroline Herschel   she whom the moon ruled ... —Adrienne Rich

flux of calculations & deductions. all night star seducing
exhausting tea making & sidereal astronomy
        collapsing horse manure
into glue to hold
        the heavens straight
a soprano waiting for the nod
to gather life's dust for hobby. to sweep some nebulae her way

she rules in notebooks for precision. *wants to find a Lady's comet*
& documents changing clusters & twin lights circling each other
        nestling swans. she is the galaxy. she is a magnet
        she is
                home. William has just discovered Uranus
                her pride expands to the swelling moments
                of the Big Bang. her eyes
dizzy at times while she cracks pencils
She feels her brother did

(*I am a mere tool ... I know how dangerous it is for women
to draw too much notice ...*)

all the sharpening

after William, she catalogues another 2,500 positions
nothing – not sleep or meals take her away
        from this proven love. for 97 years
                she emits curiosity
        for all years after, she delivers
                        a slice of the unknown

Orion

Examine closely – nebulae on a moonless night
can mimic a single blurry star. We survey through
this telescope as young bees – open stellar hives
reveal nurseries, composers creating new work
Nestled amongst Orion constellation, swinging
midway from the sword. Swimming fish. Many
cultures

decipher its ardent presence

possible black hole

in its heart

In this summer

breeze we see a

trapezium cluster

smouldering scintillation

plasma, gas, dust – ionised
atomic hydrogen, sulfur-red ribbons of orange
plush magenta in space – collage of uncontained
soaring peripheries. Inside, on its way to forming
planetesimals, its genealogy unravelling in this
nightscape. Light-corals inhabit this Via Lactea
What it means to be near, connected, curious
and yet, so far away. Molecular ancestral fire

FUSION

Cecilia Payne-Gaposchkin

*There is no joy more intense than that of coming upon a fact
that cannot be understood in terms of currently accepted ideas.*

spent time measuring ‖ absorption lines ‖ in stellar spectra

       peered through a jeweller's loupe
       and poured over data etched
       into thousands of glass plates

calculated composition of stars ‖ mostly hydrogen then helium

Harvard College Observatory in the 1920s
(paid out of the director's equipment budget)

the dean of American astronomers wrote
to say her findings were *clearly impossible*
only to confirm a few years later it was fact

hydrogen, far more prevalent in the universe
than anyone had believed ‖ a million times more

radiant bearings in patchwork night
lightest elements carrying weight

Cecilia became the first woman promoted ‖ full professor
Harvard's Faculty of Arts and Sciences ‖ department head

hydrogen ‖ monarch of chemical
substances ‖ atomic number one

this stairway of acuity
       making the bridge to new discoveries
                    luminous

Kilonova

we are detectives
we eavesdrop
  billions of years ago
  two neutron stars

circle each other
desperate and breathless
  finishing their last
pressing conversation

remnants of once intense lives
cascade into a final spiral
  until they embrace
smashing platinum

  and gold into existence
  a violent coalescence
outshining 100 billion suns
their collided mass
propagating gravitational waves

across the fabric of space
  at light speed
gamma rays detected
only a moment after

we are watching
  we are listening
we see them encompass
each other completely

with their final words
rippling through us

Perspective

i.

From this vantage, Mercury and Mars
hang parenthetical, closed sentences
while the rest of the galaxy is translucent
floating caravels in a mesmerising battalion

This hill, with its cloak of wind and solace
allows me to reach and stroke Venus
peering into the beginnings of things

You stand beside me in that tan, torn coat
as stellar showers squint in sombre, velvet sky

ii.

How large our curiosity looms, your knot-thick
hands clasp the vertigo of a volcanic ridge

These figures eminent, exclamation marks
to history. You said it's important
to see more than we're told to

Discerning light from the observatory
on Siding Spring Mountain
deciphering knowledge, perpendicular

iii.

Bob Dylan sings of an almost-hidden
moon as every note falls to the ground
perfectly re-formed. Each vowel
running its fingers over my back

anticipation of answers and comfort
Lyrics, contorting chronology
Orpheus, weeping
We talk about absolutely everything
hoarding hyperbole

We are astronomical interferometers
calculating our distance

iv.

Simone Weil said, *Truth is on this side of death*
The cat is both alive and dead
and looking out the window

Warrumbungle National Park
cradling all hope

v.

Astronomers rarely need to look up
instruments detect invisible signals

Lists of graphs, diagrams, numbers
chart the unknown, unheard, unsung

You arrive home, warm by the fire
opening your mouth to speak –

phrases, un-tamed as strands of string
possibilities opening up like a box

The Moon: Considered as a Planet, a World and a Satellite

J Nasmyth, and J Carpenter, 1874:
book held in The Portico Library, Manchester, UK

if the moon won't come to you
let us build it up crater by crater

the telescope has allowed astute
eyes on its palette and attributes

plaster models as re-enactment
geology of twenty-four Woodburytypes

close-ups of Valley of the Alps, Pico
Aristarchus and Herodotus, volcanoes

this new medium of photography
against the coal heave of cloth

physiography of ample contours
inclines of rock, contorted chasms

vast black yawning depths
summits of central cones

practicing with the back of a hand
     watching light pool in the skin
taking pictures of a shrivelled apple
contriving ranges on a shrinking globe

we are reflective cartographers
deciphering codes of terrain
noting sprays of solar eclipse
how the Sun filters through

at the moment of its occultation
by the dark limb of the moon
        we will be certain
it yields no atmosphere

we are surveyors
        an *imaginary lunar traveller*
every mountain backlit, casting shadows
while we daydream about touching dust
        a silver-margined abyss
standing on the surface

we have for you in monochrome
powdery alchemy of adoration

Mae Jemison   Hailing frequencies open

Physician/Astronaut/Chemical engineer
beginning of many careers and ready
      to launch vertical
into history
Numbers count:
      in 1992, Jemison logs
190 hours 30 minutes 23 seconds
in space on the STS-47 mission
Space Shuttle *Endeavour*
      126 circles in stable transit
dizzying in thin wings above
blue and white swirls of Earth
Mission Specialist 4 knows
she belongs here *as much as any*
speck of stardust, any comet, any planet
researching weightlessness, bone cells
threading connections and insights
Images matter:
carries with her a picture – Bessie
Coleman, aviator Queen Bess –
brings a poster of Judith Jamison
dancing along for the ride
      (performing Alvin Ailey's 'Cry'
      – the half-moon of her arm
      declaring a path beyond the void)
as a child, she watched Nichelle Nichols:
Nyota Uhura levelling orders
      on the *USS Enterprise*
Mae Jemison in low Earth orbit
creating a dynamic first of her own
the gravity of this moment
the thrill of this adventure
      opening up space
      where it wasn't before

Gravitational Lensing

Our eyes crave baths of light –
playgrounds of quivering stars
an image of a blue arc on the rim
coiling around clusters of galaxies
vivid shimmer behind you in the garden
as a torch frames your silhouette in dusk
Yearning to witness an orb's slow motion
watchful long into the balcony of night
after we see quasars at distance, distorted
we want to understand how photons travel
how dark matter halos convene over time

their complexity cushioning baryonic matter
or black holes, inferring presence from
flowing accretion disks or brightness afar
Gravity flexing the structure of spacetime
warping light's straight line – you, to me
as if a universal river pools at the sides
of invisible stone. When a large galaxy
assumes front-view, focal point – remote
siblings are magnified, arching at the sides –

strong or weak, enhancing further set
stellar families. So, if foreground mass,
background and observer are perfectly
aligned, this Einstein ring resembles
an imprint, a cereal bowl abandoned
for morning play, a concentric stain
We try to see beyond what is instantly
visible and illuminate what is known
but concealed. Our bare eyes, in cold
triangulations – astonished, but unable
to locate the weight of the cosmos –
missing out on all things we cannot see

Interior IV, 1970, **Margaret Olley**

How to sit comfortably on a chair in a painting

How to rearrange pillows and lie down
with your cool lemonade, book in hand
and glance sideways into the salty afternoon

How to display artefacts as if they wear you
How to rest your temperament on a Sunday

To write notes on your arm for later
your hair, colour of clouds

To allow day to sway on a whim
To let inside out, frames ready
to fold at any moment

DH Lawrence said scientific
knowledge killed the awe of Sun

Margaret Olley
resurrected it in
our lounge room

IGNITIONS

Vera Rubin

Some of the most astonishing discoveries
    are found on the margins

stars on edges of spiral galaxies
    moving as fast as those at the centre

everything within the Andromeda Galaxy
should be dismantling
    fleeting adrift into the expanse

but it isn't – all cocooned by dark matter halos

Rubin is told by a high school teacher to avoid
    a scientific career/ become an artist

I would prefer to stay up and watch the stars than go to sleep

She examines galactic rotation curves
    redshift in clusters
    predicted angular motion

Vera Rubin and Kent Ford deciphering
light on Ford's image tube spectrograph

Rubin's calculations: there must be at least five
to ten times as much dark matter / as ordinary matter

presence of this unmistakeable form
disguised wonder in the universe
        staring back at her

The mass of visible stars wasn't enough to hold the galaxy together

        – Rubin gets her PhD at Georgetown University
unable to study astronomy at Princeton
        as they do not accept women –

There was an extraordinary amount of matter missing

Her work confirmed by gravitational lensing
radio astronomy, cosmic microwave background

the search goes on: detectors underground
particles smashing in labs, tracing evidence

Rubin – overlooked for the ultimate prize
but observed in every fragment of this exploration

City Lights, 1952, Charles Blackman

And in a golden glass I see the dream-wished day appear – and wait. —Joy Hester

These city lights illuminating
every brushstroke between us

How fierce, the brash of black
so perfect, the relief of bay blue

You, standing either side of yourself
the day closing into an essay

Before me, a striped amphitheatre
disembodiment with built-in mirrors

streaks of old Victoria Bridge, a camel
on the skyline. Two circles, up front

hinge-less, imposing and direct
Stop signs to the end of the world

Black Tar Dream   for Dorothy Porter

only when you disappear
from your inky-dim dread

at the swoop of dark energy
escaping right into yourself

will space disclose its own
rousing untangling narratives

the universe, a perched raven
as light ricochets off your car

while imperceptible objects:
near-infrared ringlets of gas

gamma ray, ultraviolet –
unmask if closely observed

we are waves, we are particles
steering towards mortality

a quintessence of divergences
branching when you reappear

you apprehend our position
everything else is back-drop

Conservation of Momentum

I know it's a sexist thing to say, but women
aren't as good at making music as men ... —Julie Burchill

i.

– Momentum is a measurable quantity
equal to the net force acting upon it –

Suzi Quatro, my Leather Tuscadero
dented notes, fuzz chords buzz electric
bass with the thump against growling wood, her
1957 Fender Precision swinging into the stratosphere
Patti Smith, words seesawing in the dark
  mouthpiece marvel with twisty sermons:
  proto-punk, art rock, poet
Sister Rosetta Tharpe, guitar gospel gateway
her spirit sublime sound a bolt forward
Poison Ivy Rorschach, strum mechanic
echo-Theremin looping down spines
  lungs adjusting to the fresh air
Ronnie Spector, dynamic swooning lullabies
showing up the forgettable critic who wrote:
  they wear skin tight dresses revealing
  their well-shaped
  but not quite Tina Turner behinds
what she had to go through to hit the airwaves

ii.

– Action equals reaction –

The Runaways: Joan Jett & Lita Ford, plurality of cool
Aretha Franklin: Queen of Soul, surpassing sunshine
        The iconic Kims: Gordon & Deal, bass staples
        (young women practicing in the mirror just for their gait)
Christine Anu, surfing the stage as a scene stealer
Moe Tucker, be still our beating drumsticks
Salt-N-Pepa, creative hot, super smooth & utterly vivacious
Laurie Anderson, a soirée of syntax, balladeer
puppeteer of paragraph, writer of movement & verve

iii.

– With any collision occurring in an isolated system
            momentum is conserved –

        Yoko Ono carving out lines inside the winter air
Beyoncé writing long into tomorrow & outselling the world
        5.6.7.8's resonating twang, frolicking with joy
Karen Carpenter, subversive mainstream, happy home of melody
Poly Styrene, fans wearing vinyl boots out of respect
Saturday evenings, made for listening to these stars on repeat

iv.

– Momentum is equal to the mass of an object
multiplied by its velocity –

Sade, the jolt of unapologetic flair
PJ Harvey, divulging everything to the universe in A minor
Gillian Welch, crowds standing at her feet in the sweep of her lyric
    Angie Hart with her honest voice in a wave of synthetic
Hope Sandoval, the soundtrack to everyone's break up
    Chrissie Hynde making it political
    Tina Weymouth making it unusual
    Kimya Dawson making it completely unusual
& to those I've left out –
    anyone lost, bumped into the sidelines, forgotten

    Without all of you
    rock is just a pebble
    waiting to be shined

We Are One Long Conversation

We are the future and we are the past. The comma and apostrophe floating in every library of dimension.

We sit on nothing more than a tender filament: a dish on a lilo on a wave. Semaphores in the meandering grace of the universe calling to each other across the expanse. Unbounded. Sucklings in the growling nests of space. We are fervid elaborations of our best selves. *Scatterlings*. We exist as music, each note sliding into another, an exalted purr of song: rising / falling / collapsing / not-collapsing.

We break up. What if, then, you shatter into scrappy embers? What if I become a lullaby sung to young children, as stars pick up guitars and play in chorus with a D diminished seventh? What if we fold space and go back to the very moment we parted, our lips still bereft of any epitaph? We were once full of swing, luminescent constructs launching ourselves into the unknown. What if we took that walk by the river? What if we weren't heady quasars devouring all other matter? What if I wore red socks, what if you ate paella and I knew the theme to your secrets? One different neutrino and it all could be erased.

We are slits of laser light splashed on a horizontal screen. Both wave and particle. We progress further. We sense dichotomies around us intuitively as if it was as simple as sleep. Weightless. Easy. We stamp ourselves in the air like God inside thunder. We just didn't know that we *are* the string. We scan, curvatures of memory flitting about, always two places at once. We are in the cellar as much as we are in the loft. We are all centaurs, made of the eloquent slapdash of stars.

We wake to every shave of spacetime. Monumental extensions. We sail on within the nexus of this universe. It splits when action or no action is taken. We are an oscillating lattice of hope. Entangled decisions rippling. The past is a tsunami of probability. We curate opportunity.

We split. You are there. I am here. We combine. Our futures break into endless retorts and digressions. We are mariners. In one universe, I am not born, and in another I prowl all eccentricities for meaning. In one universe, this room is buttermilk, in another, some of you are missing. In one, gravity is so strong, faces stretch out beyond the week, slumping out of exhaustion. In this universe nothing can keep us back. In another, you hold the violin, and in this, the viola d'amore. Barely noticeable but somewhere an Eltham Copper lands uneasy on a mantle.

We are the tree: because of you, your daughter exists and it was the result of one moment, hanging out in the wind as effortless as sound. Obvious but not tattooed onto time for certain.

We are the endless slices of bread, one sliver of universe, placed next to another next to another, next to another. Cylindrical sheets of space intersecting membranes. Bubbles of universes pressed against each other as if one lover caressing the back of another's shirt. So far away, never risking themselves to say anything. One universe is made up of the giddy stuff of electrons but no protons. The next, lightning and perspiring angst. Another, the calm note of atoms, but no air.

Infinite. Possibilities that rehearse and rehearse. One viability smashed up against the alternative. Absorbing. The vibrations of string, resonating and pluck. You are the cello kissing the Sun. You chisel your way into being, carving out your life's curriculum. We are curled up points in every volume or sometimes, nowhere at all.

Holograms at our own birthday.

If we take the simple equation that we are all relative and non-relative. If we are both the clavichord and the bass. If we sit down at the beginning. If we just meet at the ends. If we hold on long enough to form a loop. If we are both alive and dead. If the moon closes its eyes, can we still exist?

We meet on a Saturday. We saunter by the lake. We talk orthology. We take a moment to let everything in. Every sub-atomic particle pulses. We laugh with the flush of newness. We unravel into each other. We make up the distance of quarks by reaching in for the concept of closer. We arm-wrestle doubt with pleasure. We continue along the continuum. We accept our duality. We know we are humbled by this trek. And in this universe, we don't break up.

We are one long conversation.

Portrait at Howqua River

Manna gums stand guard at midday
drooping she-oaks pep with chatter

cockatoos scuttling on our approach
flying upstream before we sit down

      our boots, patchy and brash
crave restoration while we evade sun

we drink and knit improvised garlands
      out of fallen amber flowers
lying restless under spindly trunks

we are idle in the susurration of trees
barely making it off our picnic blanket

leisurely, dipping our toes into the mud
      or splashing knee-high
a temporary lagoon amongst long grass

whirls of water, the Black Eye Galaxy
interstellar dust lanes purl the surface

someone from slightly higher ground
snaps a photo of us all, mid-sentence

this river emphatic in its lamentation
its dirt mouth dissolving into our skin

as silent stones line the shape of words
glimmering pixels of our lost afternoon

Cosmos **Revisited**

Science is a way of thinking ... more than it is a body of knowledge. —Carl Sagan

i.

Sitting with a postcard
my friend has sent me from Izmir
      flicking the corners with my thumbnail
            listening to 'Mood Indigo'
      whipping up theories
      on the beginnings of life

I think of myself
      as I did when I was a child
      without Earth, without space
         without time

ii.

Books become a communal memory
not stored in our genes or in our brains

submerging inside my worn copy of *Cosmos*
configurations of curiosity & science

this postcard talks of Byzantine emperors
museums, rivers & ridges to the south

the words in both: stellar literary records
long threads of history held in my hands

a feeling of cosmic loneliness
                              & galactic togetherness

iii.

Time & space are fused:
      it takes around seventy-two years
for the light of Mu Cygni to reach Earth
      each time we scope this binary star system
      we are looking at it
when Hubble was peering through the Hale Telescope
saying: *hope to find something we had not expected*
the physics world was celebrating work on cosmic
rays
        & my grandparents
        had barely known one another
still too shy to ask each other out to a dance

iv.

All things are relative:
a time before nuclear power
   a time before industry
   before libraries, before language

when we were governed by instinct
where we lived by sound & rawness
when we feared storms
when we revered nature
when we lived by cycles
a time of survival
when we couldn't find a lodestar
when we didn't know how to name it
  before the neo-cortex, the limbic system
    in the early days of the R-complex
  before we could taste & smell
  when we were single-cells
before sex was invented
before rock was formed
  before planets cooled
   before matter
    a time before time

v.

We bond together in *starstuff*

> with sit-coms, war, particle accelerators
> physics problems, pimples before a date
> politics, famine, cause & effect, surfing
> cryogenics, gene manipulation, buildings
> backgammon, phone calls, parents, empathy
> music, postcards & apple pie from scratch

as Carl Sagan said:

> *these are just a few of the things*
> *hydrogen atoms do*
> *given*
> *fifteen billion years of evolution*

Physics of Mourning

We live in a world of unfolding and becoming. —John Polkinhorne

Time is only a process
      not physically positioned
      in space or in this room
I massaged your aching feet
the lilies spill into the light
Einstein's theory of special relativity
confirms time slows or speeds
depending on how fast you move –
comparative to everything else. I am
completely still, you have disappeared
Time has duration. Our conversations
were endless. They began before they started
      one night you pulled me closer
      to hear your words on how hope
replenished your universe. The magnitude
of your kindness elevating the new moon
Frames of reference help the observer
measure an event. I remember how you
liked your coffee, how your face would tilt
towards the Sun but your hands are lost to me
as if they just fell away at the bottom of a page
The dog doesn't know where to sleep
no-one knows where you hid your letters
Gravity forcing me to sit quietly as I try
not to collapse. This weight of grief –
a trillion goodbyes at once. Time is not
the barrier, it is only the conduit
in which your memory travels back
a clock face, windowless without expression
      affirming you are not here
      but were everywhere, once
Your yesterdays behind you –
now extend in front of me

A Kind of Humming Silence   after Rebecca Solnit

Every day is weekend now – our backyard the only
connection with outside. We cannot venture beyond
        this perimeter of blanched ivy vining
        its way down brick or stray past the boundary
of whisky grass weeping at the front porch. Topography
here is convened into main areas, co-ordinates of terrain
        in need of navigation – creating continents
in the dirt. Not only to make time drift, as if becalmed spinnakers
but to feel part of something, to breathe in the garden
with its imaginary flat plains, forests, ridges and valleys

You stand by the side fence with its curled rock pitted
in the corner, now pretending to be Algar de Benagil
the sea cave on Algarve Coast, Portugal. Watching you
snoop through gaps – reds bleeding into concrete – I picture
        you in surf, holidaying far from the lawnmower
        whirring. At your feet are stones but I daydream
barnacles, ready to withdraw into their shells
        white cones in the shallows of your toes

These spaceless moments remind me – sometimes
we are many places at once. Anywhere but here in solitude –
    other times, the eucalypt casts shade at exactly the right angle
    black ants scurry in wattle around pollen, as the Sun
butterscotch-fringed and sharp, runs its spurs along my neck

From here the backyard is vast – exactly where I want to be

You, with your unruly hair, me in a sunken chair with a newspaper
     If I tilt my head, the landscape becomes oranges
     and yellows, Grace Cossington Smith's *Door into the Garden*
appears – how the greens fold into brown – squares
of light drop at my feet, carpeting my way to you

This is our whole world. Our resolve as thin as crêpe paper
our skis, dusty and our bathers, dry. Hope, broken-limbed
                    We've camped out here, bought binoculars online
to observe the micro-universe of insects. We've made spires out
of table legs. We've listened in on the quiet long enough to know
that quiet is never there. Even in the murmur of tranquillity
                    somewhere a Jack Russell will bark
                    This backyard needs its own atlas, to record
every bump, every lattice of fern, every tuft, every lemon

All latitudes are mapped – paces I have taken
                    shrubs you have watered, posts we have
mended. Every longitude marked – topsoil we have nurtured
                    doves we have fed, each star

We will again enfold as the day closes, eat by the fire-pit
this time, eyeing the Milky Way as it pours
into the enormity of space, the galaxies – a nod
          to the multitude and fullness of nature below
You and I sit in the hush of warmth
quietly waiting for escape

RED GIANTS

Carolyn Beatrice Parker

It is a silvery metal
      in a dark room
blue-skewed glow
excited by decay
Does Parker
      hold the polonium in her hands
does she ever breathe it in?
Working government top-secret
– The Dayton Project
part of the Manhattan –
research and development
Radioactive reverberations
kindling for a new world
Employees not allowed
to eat in processing areas
      scrubbing down before they leave
      (some have contaminated bobby pins)
Parker, first Black woman in the U.S.A.
      with a postgraduate
      degree in physics
two masters – the other, mathematics
dedicated, hardworking
Assistant professor
      in physics at Fisk University
close to completing a doctorate at MIT
      afire
microscopic amounts known of her work
within this team constructing secrets
strikingly more to discover
from this bold ascending scientist
      her time far too short

atomic number 84
leukaemia age 48

Interpreting Red

I'm conversing Rothko in whispers
lifting my fingers, exalted, cathedral-like
coaxing my friend who knows the craft
of frozen breakfasts, pool table scuffles
& lounge room posters & reading
everything as if it were a drunk manuscript
Now he stands before this ache of colour
in fluster, blushing deferential-cranberry
Is this carnal? he asks with whiplash
Cowlick whorls diffuse his question
as we crumble into its distillation
scorching with fury & dissonance
He imagines a stolen waltz, I dream
the volatile margins of a goodbye letter
We both see the rumblings of life
& then – moody particles dissipate
leaving the Sun to fend for itself

Robes of Darkest Blue

After receiving the diagnosis —
such rapidity, compulsion, permanency
my body, a series of isochronous vibrations
there's no way I can hold hands with fate, so

you stand at the edge of waves — your frame
a cratered shipwreck as foam & water puckers
around the trim of your jeans. Everything
about the light is granular, even the gulls

are overcast. You say: a marine animal's whole
world is sound. I know what you mean
Depths so chasmic the light can't reach them
beds of oceans, absent of sun but never

singular. Below, a procession of wildness
& richness — hydrothermal vents, volcanoes
vast canyons — a plethora of growth. Sonar
communication is full of bustle. I say: *this is*

how our darkness is embraced as eddy, rushed
sequences — rhythms of uncertainty, tread in
a chit-chat of tide. Wind whips up your hair
as it covers your face. Sand, fibrous & husky

ruffles against our skin. You sigh with a roar
of a winter afternoon, lungs drawing in every
force they can muster. Boats – yo-yoing up
& down, bereft of passengers, precarious

You pause as if this is your fault, but here's
the thing: we are more than sadness falling
into the tip of the sea, more than cuts of icy
wail tumbling at our feet. You – in a jumper

with holes & memories of loss will make it out:
inspect rugged crests, jump onto passing ships
glide over storms, pick up seashells, crab-suits
without muscle or old bottle tops. Palpitating

emotions skipping like stones, on waves rising
& falling with each breath & if they don't go far
know – the way you ride each passing current
you can. With all your tendrils of rebellious life

& this untamed reef, nestling on the brim of time
        distilled in a daring canopy of space
will outlive every last one of us

Constellation Rifts

i. Collisions

I don't even drink coffee but I was told over the phone, *look, you have cancer, can you come in?* I was standing in line at our café, so forever I'll remember the rapid pinball talk of workers manoeuvring through conversations and wit – coffee-bean aroma punctuating the air. I can see my hand pay for my blueberry muffin. *Of course, I can be there in twenty.* Wasn't I supposed to be sitting down? Not waiting in a space I often ate my sautéed mushrooms on toast.

ii. MRI

My head face down as the MRI is clanking away. Trying to count to one-hundred, or see myself floating in space – looking down at an Earthrise or imagine the Horsehead Nebula with its delicate folds of gas, cantering across space. But I think of Jocelyn Bell Burnell using radio waves to discover pulsars, how she interpreted composition, structure and motion. These magnetic fields and waves now assessing me. My breast images are lit up with jets from hypernova. At least, that's what I deduce.

iii. Scintimammography

After the injection, laying undressed, jagged – brittle but crisp and alert and the pain – twenty bees scurrying as if startled – stinging the nipple over and over and over. Weight radiates from chest to arm, from tender skin to lymph, thermal and liquescent sensations until everything cools and settles, zipped into the mind. Metallic mouth. Heaviness releases so the neck can move slightly – an aerial view focuses on contrasting screens of shadow-black and faux-phosphorescence. These lucent mini-supernovae, vital and living, lantern dancers forming galaxies. The radiotracer, emitting gamma rays and swaddling tissue, grading as it details function, probing for signs of trouble. Gamma rays, detected after collisions of two neutron stars – regular from cancer, old self and new, tranquil and unquiet. Nothing is as before. This will guide the surgeon – excavating cuts, new stitches, mortal scars. An aperture, liminal lamps into darkness.

iv. Surgery

Debris of cuts. Pieces missing. Flattened and tied up. My skin abrading. I'm shedding. Displaced. On heavy medication, I can't manage thumbs to text friends but I order two pairs of zebra print shoes in the right size. I think a ship has passed my window on the second floor. All memories ripple, one universe, sideswipes another. Relief, as colossal as Jupiter.

v. Radiotherapy

Breath hold. So, they don't radiate your heart. Strapped in with arms set above. Micro movements so every angle is exact. Each morning at 8am, Changing Room A. Same room, same gown, same footstep. Erasure, evanescence. High-energy X-rays, protons and particles passing through with the buzz of stop/start/stop/start. A labyrinth waiting to finish. The metrical beat of marking a calendar. Hold. Breath. Release.

vi. After

How long is after? How far, the future? I dash to a boat then halt above cavernous Phthalo blue ocean, anticipating undisturbed moments, catching sight of the saddle-tail snapper. Irrepressible winds disrobing our hair. All the while laughing. Some days are that. Some, just sitting in the kitchen with a book. Our humbling molten existence. Our eyes, full of the Rosette Nebula. Our minds, often in the needlework of living, hurtling towards an inevitable end.

Migraine after 'Blue, Orange, Red', 1961, Mark Rothko

Each planet spools around the Sun in
curves — foreshortened circles — ellipse
iridescent trails saturating my skyline

this ramble of red smudging the night
baritones of blue penetrating edges
orange submitting, grounding everything

Rothko assembling windows in my brain
every frame, an amplitude of oscillation
an imagined Moiré pattern coruscating

as each brushstroke ruptures repose
Here — time is unsnarled in sequences
scintillations tinging my eyes for the day

subharmonic voices simmer, melismatic
In this proliferate architecture, I'm
searching everywhere for a balustrade

needing to calm myself and close down
resonating hues are piercing — thought
and conversations arrive in reverse:

in darkness, nebula and meteors meld
my words — barely hanging in the air

Carrying Black Holes

For four years Stephen Hawking's
speeches rustle in my bag, his pocket
                BBC Reith Lectures
suspended in time – dispersed amongst
Japanese brush pens, miniature staplers
spare violet sunglasses, polychrome scarf
and a journal with asteroids on the cover
My pencil summaries in the margins:

Pairs of virtual particles could fall in
Laplace said if we know the universe now
      we will understand its past –

moments I've laid out a blanket
or read it on a train ride home

inspiring me   to consume

more   knowledge

reading   how

the event   horizon

             draws light back in
coaxing swiftly, not allowing it to flee
quasars twirl around boundaries
information hidden from everything else
(maybe not lost
only on its way to being reframed)
Hawking's assertions on each page
– lines packaging dense space
carrying the mass of these facts
with me everywhere closely, heavy
never allowing escape

JW Space Telescope Stirs

how we tell ourselves the bleak expanse
of nothingness is something | observe

calibrating the past | spatial increments
lambent assemblies of branched history

bold telescope in halo orbit | L2 Lagrange
point | waiting for clarity in staring cold

its primary eyes scanning | 18 hexagonal
mirror segments | gold-plated beryllium

scouring galactic light | deep field
nebula more than sparkle on a poster

how we keep searching | things larger
than new space | we are infinitesimal

honeycombs surveying long-wavelengths:
visible glints | through to mid-infrared

anticipating the mumblings of first stars
life's breakfast | formation of planets

during alignment process | self-reflection
precision composure | resplendent focus

discovering where we have been | where
we will be | a prime lens on our universe

Heat Death

space
      stretches
            space

as gravity wrestles expansion
we are stories of rotations:
fluctuating oceans of atoms
shawls of stars gathering

in pearly-lavish galaxies
dark matter embracing
as dark energy hastens

the universe – 13.8 billion years
its radius 46.5 billion light-years –

     e x t e n d i n g

one day, everything will dissipate
frayed fingertips unable to touch
no cannonades of new sparks
forming – a fugue of entropy
this thermal equilibrium
the coda of a burnt fuse

heat – disordered energy
harbinger of eternal cold
furrows of trembling matter
blurring further and further
into the spreading vacuum

     slow sleep
        of disappearing
          tombs

Sagittarius A*   Event Horizon Telescope, May 12, 2022

how we wanted to know
what held firm at the centre

a sense of generational longing
only cured by operatic narrative

supermassive black hole shadow
glowing ring with cavity puncture

from far: brooding, slumberous
closer: orange-red smouldering

fast spinning silhouette wreath
devouring infalling surrounds

gas, debris swirling its perimeter
as stars slingshot around the rim

the interstices between itself
and those who view back home

an undressing blur of wonder
four million times that of our sun

synced observatories, collaboration
how we come together, uncovering

an image, years in lucid construction
calculations unveiling visions of data

core of our Milky Way hub singing
as we assemble the transit of notes

SUPERNOVA

Jocelyn Bell Burnell

If we assume we've arrived:
we stop searching, we stop developing.

proud pulsars
residue
of massive stars
      collapsing into
      a neutron star
strobing supernova
      spherical and dense
      size of a bustling city
      more mass than our Sun

beams of electromagnetic radiation
whirling from its magnetic poles
      recurring signals
      palpitating refrains from space
– first observed in 1967
by Jocelyn Bell Burnell
and Antony Hewish –
Someone takes a photo:
Look happy dear, you've just made a discovery!

millisecond pulsars can siphon matter
and momentum from their companions

in 1974 Hewish receives
a Nobel Prize in Physics
but not Bell Burnell
she wins other weighty awards
becomes a Dame, donates
substantial sums to further
those ignored in science

pulsars radiating light
in a manifold of wavelengths

On Listening to the Astrophysicist

She hears in numbers. Scribbles in zeros
plump, exhausted zeros, deft at first inspection
      then you blink & they don't quite finish
          oblique gravity opening
letting layers of the universe escape through one side
All her calculations lean to the exit, towards
the door. You are a swamp of desperation & charge
ready to transmute into a single figure:

a cracked script of a forgotten language
Her theories are a Mondrian
fragmenting perception with geometric retinal imprints
if you assimilate everything she has spoken, you will
          simply not exist
You have been advised:

don't go out with a genius, they never sleep

Her voice, raspy, the perfect afternoon
you sift through your memories for any leaks, portals
into another time to see if you ever had a chance
of becoming an astronomer, through tapestry of chaos
          (probably not)
but you want her formulae chalked into your ribs
Light bulbs simmer above, the zenith
of this lecture hall. Her ideas resonate heat
pulsating to the finger-tips of your notes

Views of the Architecture of the Heavens in a Series of Letters to a Lady

JP Nichol, 1837: book held in
The Portico Library, Manchester, UK

Madam –

Dear Public,

I cannot deliver you all of astronomy
only paint awe and sum of magnitude

motions of clusters in possible infinities
formations and swathes of uneven space
our firmament, the entire mass of stars

Trace your finger around the line
of the Milky Way as it branches in two
the shape and dimensions of this swarm
elongated as outlines finally dissolve
a *diffused starriness* in the ribs of galaxies

intervals between each flashing kernel
hollow-black, *external and obscure vacancies*

Are the different suns isolated or related?
        Patterned *from the womb of nebulae*
their effulgence softening in the distance?

We study with the power of new refractors
      drawing and harvesting upcoming
charting boundaries, pinpointing radiance

filmy or *nebulous* fluid shining by itself
      endless diversities of character and contour

Here on Earth, we are modest
The Great Book of the Universe
fathoms more in comparison
– this book, then, *must seem sibylline, incoherent*
      but we are not fragmentary

I cannot deliver you all of astronomy
I can only hope to detail and share –
immensities, glister or phenomena
      and unity of all celestial things

        Yours, always, in science, etc.

Polar Night

Last flights whisker out in February
while Antarctica retreats from the Sun
Science winterers settle in as its heliacal
flare dwindles, midnight melting twilight
To the eye – this season swells the mass
of the continent, cold fingers on a map

Transantarctic Mountains partition west
from east, imposing, steep underneath
vast ice shelves, floating summits of freeze
craggy-white rafts of towers fed by glaciers

Presses of emperor penguins are breeding
on sea ice. Huddled male bunches in warm
circles, taking turns on the outside, female
posses, diving for krill in briny thickness

Stratospheric polar vortex shifts fiercely
shaping sastrugi, riding the peaks of wind
Those who stay replace sunscreen for GPS
units and headlamps, careful not to venture
too far. Scientists complete observations
calculating, maintaining, thriving all year

to break up work, they take in the Aurora
Australis – charged particles from solar winds
jostling their way through the magnetosphere
– feathery purples or pinks, shimmery greens
splintering an elongated scattered night, until
finally, summer collects in the aching horizon

Higgs Field

Why we don't disappear into the noiseless frailty of existence.
How subatomic particles navigate this field in distinct ways.
Energy pervading. Top quarks with more mass than the electron.
As if electrons slide slick – marbles on marble, glassy rolls
smoothing through brisk expanses of space. Polished glissades.
The top quark, a pinball, trundling along thick pastures. Varying
resistance. Photons, neutrinos, seamless gliding in the breadth
of the known-universe unable to stop or mingle, impervious
to surrounds. Fundamental particles attain mass – consequential
interactions – ongoing discussions in magnitude. Excitations
creating the Higgs boson, a kindling of impression lasting
1.56×10^{-22} seconds. How we converge into bundles. What we
find in the decay. How we observe only after. This scalar quantum
field of transformation essential. How we hold together. How
we might understand gravity. Why we can witness sinuous stars
palpitating on a winter's night. Why we embrace as we fixate
upwards, breathless, at the barred smears of Andromeda and wait
just a little longer to go home.

Mysteries of Time and Space

Richard A Proctor, 1883:
book held in The Portico Library, Manchester, UK

What is there beyond the starry vault? —Louis Pasteur

Even in the epoch of Tycho Brahe
we believed stars were anchored

but everything is in constant motion

one day the moon will pack up its things
receding into withdrawn sheets of space
        (fractured off from the Earth and assumed
        the dignity of an independent body)
this sleepy airless satellite
will again retire from any binds

Moon *she* / Sun *he* / Earth *ours*

serrated Sun with its orb-life lustre
sways the planets by its attraction
holding no perpetual energy
        mortal and resplendent
in full eclipse with its streamers
a white halo and shining nimbus
        the corona is a true solar appendage

tenuity of comets leave cold hazy trails
     electric-oid action of some kind
not portents of catastrophe but keys to knowledge

Mars in transit reflects delineations
nine of the seas ... have this peculiar shape ... bottle-necked

these five stages of any world's life:
     lustrous vapour, fiery youth
creation-bearing middle age, degrees
of decrepitude, then, ultimate death

     the universe as we know it, tends to an end –
which may be the beginning of new forms of existence

when our home finally resigns
     radiating iron will swelter
clouds slouching in thick atmosphere

a measured rush to the finish
unavoidable narratives of time

Heisenberg's Uncertainty Principle

The reality we can put into words is never reality itself.

we cannot measure
position (x) and momentum (p)
of a particle with absolute precision

as the chances of predicting position rise
the prospect of knowing momentum falls

waves ~ disturbances extending into space

if a wave with a measurable position
        collapses onto a single point
        on an indeterminate wavelength
it will have indefinite momentum

if a wave with measurable momentum
has a wavelength oscillating infinitely
it will have an indefinite position

binding interconnection
        conjugate variables:
energy/time/velocity/momentum/mass
linked relationships

we can determine your speed
when you run your fingers through your hair
                on the long drive home

but not the pressure of guitar calluses
as you grip the wheel of a rusty-hulled car

our everyday life ~ a supercluster
not Planck's constant

if you are the electron ~

you can either know where you are
or where you are going but not both

such small scales
distilled action

you are here
even though you have gone

Cloud Chamber

Not the particles
but where they have been
Radon atoms with short deep trails
muons arrowing unimpeded
electrons kinked and twisting
or unswerving depending on energy
positrons excited by the magnetic field
      protons dotting or streaking
      across the expansive black

Cloud tracks differing in shape/length/width
alpha particle/ beta particle/ gamma ray
some resembling untamed fireworks or
sea creatures curling on the ocean floor
      magma rippling over and under caves
      lines splashing spry into the landscape

Wisps, sweeps of paths
Ions positively charged
and vapour – negatively so
      attracted –
      as it condenses
      exquisitely onto the ion
Radiation emissions unfold
as extroverted chronicles emerge

Watching decay
life affirming
how we will all
      one day
         transform

  into
something else

Bose-Einstein Condensate

How we look at superfluidity and superconductivity.

states of matter
within a shiver
of absolute zero
atoms slow

drawn-out
languid
so cold
they band

together
every particle
at once
everywhere

wave packets
swaying
elongating
embellishing

bosons
losing identity
flow forming
overlapping

decoherence
where one ends
another begins
within a quake
we are

(listen
(the closer we are
(we crumble into
(black pools

(we become one
(scattering into
(time-thin
(mantles cooling

(on the surface
(we pulverise
(into embers
(teaspoonful

(by teaspoonful
(you are fibres
(sleepless
(we are soaring

(into the thick
(of things
(listen
(to the numbing

(dark
(learn from
(this wave-function
(cold matters
(at rest

Spectrum Analysis in its Applications to Terrestrial Substances: and the Physical Constitution of the Heavenly Bodies

Heinrich Schellen, translated and revised
by Jane and Caroline Lassell, 1872:
book held in The Portico Library, Manchester, UK

Light is sleek and wordless, filling intervals
pores of space, travelling without impediment

disseminating in the body of the universe –
an immeasurable sea *of highly attenuated matter*

imperceptible to the senses. *Although the theory
of light is now so completely understood* there are

many ways to clearly see in this ether, this spectra
not ghostly apparition but all colours in a woven

prism. Belts of absorption lines, marks where
it is restrained, absent: *luminous vapours* in shades

revealing grain and texture. We view the voltaic
arc spiking an *electric spark* between metal poles

in the stratum of air. Foucault's galvanic lamp:
currents lacing gaps between two end-to-end

carbon rods or brightness as the Bunsen battery
produces much discomfort to the eyes, elements

touching so their ends glitter in blaze, ego-hot
as the *electricities* attract. *Geissler and Plücker's tubes*

or rainbows in rectangular parallelepiped bars
of glass. Limelight of the flickering stage, calcium

oxide burning, a caged opera of glimmer, noticing
speckles of planetary nebula or bursts of splendour

with *gas-streams in the sun, balls of fire seen through
a telescope*, importance of illuminating our past

what insight we will burn into the future

Anticipation of Light

Second image released from M87
Event Horizon Telescope, 24 March 2021

Stare at this too long
the aperture of blackness
        opens up, leaving
the core of the image
expanding towards you
at the speed of a blink
Supermassive black hole
55 million light-years away
in the Virgo cluster
6.5 billion times more
        colossal than our Sun
seen here in polarised light
        a pavilion of swirl
brushstrokes burnt yellow red
elucidating spacetime's bend
magnetic field lines etched
at the rim of the innermost
        stable circular orbit
This region of reverence
        named Pōwehi
Hawai'ian for *adorned*
fathomless dark creation
        jets of energy extend
5,000 light-years from its centre
communicating in the entire
        electromagnetic spectrum
A new view edging us closer
to understanding the singularity
M87's structure and behaviour
but until then —
we gaze long into the abyss
and wait

Emmy Noether

My methods are really methods of working and thinking;
this is why they have crept in everywhere anonymously.

No matter how hard we try
energy cannot be created or destroyed

    Noether's abstract algebra
    still radiating insights to this day

Noether's theorem states:
every differentiable symmetry has
a corresponding conservation law

    her mastery incandescent but
    no salary until she reached forty
    equal but unequal to some

Her numerical and brilliant fingerprints
etched into so many branches

    Noetherian: group/induction/problem
            ring/module/space/scheme

The mathematics a foundation
for the standard model of physics

    changed the face of algebra
    absolute beyond comparison

Laws of physics remain the same
as we move around the universe

but for Noether they bend more deeply

    as her legacy exponentially increases

ACKNOWLEDGEMENTS

Galactic thanks and deep appreciation to Kent MacCarter and Cordite Books. Endless thank yous to Nat Bates, Gayelene Carbis, Nathan Curnow, Trisha Dearborn, Professor Alan Duffy, Adam Ford, Nicole Hayes, Michael Leach, Dr Katie Mack, Natasha Mitchell, Paul Mitchell, Lisa Gorton, Sar Ruddenklau and Andrew Watson. Multiversal love to Jordie Albiston, whose support meant everything to me, and to Andrea Rassell for her generous introduction and advice.

Thanks to all the scientists, scientific communities and institutions for sharing their passion, wisdom and kindness with me. I am forever in awe.

Poems in this collection have previously appeared in *The Age, Aniko Press, Australian Poetry Journal, Australian Book Review Podcast, Baby Teeth Journal, Best Australian Poems 2023, Best Australian Poems, Best Australian Science Writing, The Blue Nib, Cardenal, Cathexis, Consilience Journal, Cordite Poetry Review,* Science Gallery Melbourne's *Dark Matters, Dark Sky Dreamings: an Inland Skywriters Anthology, For Women Who Roar, Going Down Swinging, Graviton, Hecate, High Shelf Press, Intima: A Journal of Narrative Medicine, Meanjin, Outer Space, Inner Minds, Rabbit Poetry Journal,* Science Museum Group website, *Science Write Now, Seisma Magazine, Stilts Journal, Southerly, Tamarind, Teesta Review, Westerly* and *The York Literary Review.*

'A Kind of Humming Silence' was highly commended in the 2021 Bruce Dawe National Poetry Prize and 2023 Poetry Kit International Poetry Competition. Part of 'Constellation Rifts' was shortlisted for the Ada Cambridge Poetry Prize, 2023. 'Perspective' was winner of the My Brother Jack Poetry Award, 2014 (previously titled 'Universality'). 'Gravitational Lensing' was a finalist for the Raw Science Film Festival, SCINEMA Film Festival, Mannheim Arts and Film Festival and award winner for the Poetry Film Festival. It was also selected for the SciFilmIt Science-Film Competition and the Nature & Culture Film Festival. 'Gravitational Lensing', 'First Three Minutes', 'Kilonova' and 'We Are One Long Conversation' were aired on Radio National's *Science Friction*. 'Heisenberg's Uncertainty Principle' has been displayed in the Deutsches Museum, Science Communication Lab, Munich, Germany (in collaboration with *Consilience Journal*). 'Black Tar Dream' came out of the ANAT Synapse Residency, 2023. With special thanks to ANAT and Professor Tamara Davis and all her team at University of Queensland's School of Mathematics and Physics.

Some of these poems featured in the planetarium shows *Elemental* and *Particle/Wave*, with special thanks to ARC Centre of Excellence for Gravitational Wave Discovery (OzGrav), Museums Victoria and Melbourne Planetarium. Segments of 'Black Tar Dream', 'Heat Death', 'Vera Rubin' and 'Higgs Field' feature in the installation, 'In This Room. Everywhere' (Sometimes/Watson) as part of the *Dark Matters* exhibition for Science Gallery Melbourne, 2023. With special thanks to Science Gallery Melbourne, Arts at CERN and the ARC Centre of Excellence for Dark Matter Particle Physics. Some poems were also written for the Manchester City of Literature and Manchester Literature Festival virtual writer residency, 2021. Special thanks to The Portico Library.

Many of these poems were written with support from Creative Victoria's Sustaining Creative Workers Grant and the City of Melbourne's Boyd Garret Residency.

Australian Government

Alicia Sometimes is an Australian poet, director and broadcaster. She has performed her spoken word and poetry at venues, festivals and events around the world. Her poems have been published in *The Age, Best Australian Science Writing, Best Australian Poems, Griffith Review, Meanjin, Westerly* and more. She is director/co-writer of the art/science planetarium shows, *Elemental* and *Particle/Wave*. She has been awarded residencies at Katharine Susannah Prichard Writers' Centre, Varuna, the Melbourne Aquarium and was a Creative Fellow at the State Library of Victoria. In 2021 she completed the Boyd Garret residency for the City of Melbourne and a Virtual Writer in Residency for Manchester City of Literature and Manchester Literature Festival. In 2023 she received ANAT's Synapse Artist Residency and co-created an art installation for Science Gallery Melbourne's exhibition, *Dark Matters.*